Disclaimer

The publisher of this book is by no way associated with the National Institute of Standards and Technology (NIST). The NIST did not publish this book. It was published by 50 page publications under the public domain license.

50 Page Publications.

Book Title: Advanced Engineering Environments for Small Manufacturing Enterprises: Volume II

Book Author: Steven J. Fenves; Ram D. Sriram; Young Choi; J P. Elm; J E. Robert

Book Abstract: To assist the Small Manufacturing Enterprise (SME) in adopting Advanced Engineering Environments (AEEs), this report provides two self-assessment tools; the Self Assessment Tool for Engineering Environments (SAT-EE) to assist an SME in assessing the adequacy of the current computing support environment in handling technical tasks, and the Self Assessment Tool for Engineering Capabilities (SAT-ETC) to collect the needs and desires of the operation, and map them to the capabilities of specific classes of CAD and CAE tools. These tools help the SME evolve from lower AEE levels to higher ones, shifting operational modes and philosophies to gain the full benefit of the AEE. In choosing an AEE component the SME must consider the tool?s functional capabilities, interoperability, usability, and expandability. Selection, procurement, and installation of a new technology must be followed by tool adoption to integrate it into the SMEs operation. A six-step process based upon the Shewart Cycle has been used successfully to introduce advanced technologies into SMEs.

Citation: NIST Interagency/Internal Report (NISTIR) - 7090

Keyword: Advanced Engineering Environments (AEEs);design;Engineering Capabilities;interoperability;SAT-ETC;Self Assessment Tool for Engineering Env;Small Manufacturing Enterprise (SME);tool

NISTIR 7090

Advanced Engineering Environments for Small Manufacturing Enterprises: Volume II

Steven J. Fenves
Ram D. Sriram
Young Choi
Joseph P. Elm
John E. Robert

National Institute of Standards and Technology
Technology Administration, U.S. Department of Commerce

NISTIR 7090

Advanced Engineering Environments for Small Manufacturing Enterprises: Volume II

Steven J. Fenves
  National Institute of Standards and Technology; Manufacturing Engineering Laboratory

Ram D. Sriram
  National Institute of Standards and Technology; Manufacturing Engineering Laboratory

Young Choi
  National Institute of Standards and Technology; Manufacturing Engineering Laboratory

Joseph P. Elm
  Software Engineering Institute; Carnegie Mellon University

John E. Robert
  Software Engineering Institute; Carnegie Mellon University

January 2004

U.S. DEPARTMENT OF COMMERCE
Donald L. Evans, Secretary
TECHNOLOGY ADMINISTRATION
Phillip J. Bond, Under Secretary of Commerce for Technology
NATIONAL INSTITUTE OF STANDARDS AND TECHNOLOGY
Arden L. Bement, Jr., Director

**Carnegie Mellon
Software Engineering Institute**

Pittsburgh, PA 15213-3890

Advanced Engineering Environments for Small Manufacturing Enterprises: Volume II

CMU/SEI-2003-TR-XXX
ESC-TR-2003-XXX
NIST XXX

Steven J. Fenves (NIST)
Ram D. Sriram (NIST)
Young Choi (NIST)
Joseph P. Elm (SEI)
John E. Robert (SEI)

November 2003

Technology Insertion Demonstration & Evaluation (TIDE) Program

Unlimited distribution subject to the copyright.

This report was prepared for the

SEI Joint Program Office
HQ ESC/DIB
5 Eglin Street
Hanscom AFB, MA 01731-2116

The ideas and findings in this report should not be construed as an official DoD position. It is published in the interest of scientific and technical information exchange.

FOR THE COMMANDER

Norton L. Compton, Lt Col, USAF
SEI Joint Program Office

This work is sponsored by the U.S. Department of Defense, and was performed jointly by Carnegie Mellon University (Carnegie Mellon) through its Software Engineering Institute and the National Institute of Standards and Technology. . The Software Engineering Institute is a federally funded research and development center sponsored by the U.S. Department of Defense.

Copyright 2003 by Carnegie Mellon University.

NO WARRANTY

THIS CARNEGIE MELLON UNIVERSITY AND SOFTWARE ENGINEERING INSTITUTE MATERIAL IS FURNISHED ON AN "AS-IS" BASIS. CARNEGIE MELLON UNIVERSITY MAKES NO WARRANTIES OF ANY KIND, EITHER EXPRESSED OR IMPLIED, AS TO ANY MATTER INCLUDING, BUT NOT LIMITED TO, WARRANTY OF FITNESS FOR PURPOSE OR MERCHANTABILITY, EXCLUSIVITY, OR RESULTS OBTAINED FROM USE OF THE MATERIAL. CARNEGIE MELLON UNIVERSITY DOES NOT MAKE ANY WARRANTY OF ANY KIND WITH RESPECT TO FREEDOM FROM PATENT, TRADEMARK, OR COPYRIGHT INFRINGEMENT.

Use of any trademarks in this report is not intended in any way to infringe on the rights of the trademark holder.

Internal use. Permission to reproduce this document and to prepare derivative works from this document for internal use is granted, provided the copyright and "No Warranty" statements are included with all reproductions and derivative works.

External use. Requests for permission to reproduce this document or prepare derivative works of this document for external and commercial use should be addressed to the SEI Licensing Agent.

This work was created in the performance of Federal Government Contract Number F19628-00-C-0003 with Carnegie Mellon University for the operation of the Software Engineering Institute, a federally funded research and development center. The Government of the United States has a royalty-free government-purpose license to use, duplicate, or disclose the work, in whole or in part and in any manner, and to have or permit others to do so, for government purposes pursuant to the copyright license under the clause at 252.227-7013.

NIST disclaimer: No approval or endorsement of any commercial product by the National Institute of Standards and Technology or by Carnegie Mellon University is intended or implied. Certain commercial equipment, instruments, or materials are identified in this report in order to facilitate better understanding. Such identification does not imply recommendations or endorsement by the National Institute of Standards and Technology or by Carnegie Mellon University, nor does it imply the materials or equipment identified are necessarily the best available for the purpose. Note that most products mentioned in this report are either trademarked or registered. The URLs mentioned in the report are current as of the time of publication of this report.

For information about purchasing paper copies of SEI reports, please visit the publications portion of our Web site (http://www.sei.cmu.edu/publications/pubweb.html).

Table of Contents

1 **Introduction** 1
   1.1   Context of report 1
   1.2   Purpose 1

2 **Self-Assessment of AEE Requirements** 2
   2.1   Self-Assessment Tool for Engineering Environments 2
       2.1.1   Introduction 2
       2.1.2   Function of the tool 5
       2.1.3   Use of the tool 6
       2.1.4   Interpretation of Results 6
   2.2   Self Assessment Tool for Engineering Tool Capabilities 7
       2.2.1   Introduction 7
       2.2.2   Function of the tool 8
       2.2.3   Use of the tool 10
       2.2.4   Interpretation of Results 10

3 **Migration** 12
   3.1   From scratch to CAD for drafting 13
   3.2   From CAD for drafting to CAD and external analyses 14
   3.3   From CAD and external analyses to a Basic AEE 17
   3.4   From Basic AEE to Intermediate AEE 18
   3.5   Beyond an Intermediate AEE 19
   3.6   Additional migration considerations 20
       3.6.1   Downstream (CAD/CAM) integration 21
       3.6.2   PDM adoption 22
   3.7   Summary 23

4 **Tool Selection** 24
   4.1   Selection criteria 24
       4.1.1   Technical criteria 24
       4.1.2   Service criteria 28
   4.2   Selection resources 29
   4.3   Summary 31

5 AEE Technology Adoption..32
5.1 The Adoption Challenge..32
5.2 The Technology Adoption Process..34
5.2.1 Understand existing company environment...............................37
5.2.2 Establish Technology Adoption Project Goals & Metrics............38
5.2.3 Evaluate Technology Options..39
5.2.4 Obtain Technology...41
5.2.5 Implement & Adopt Technology..42
5.2.6 Analyze & Deliver Adoption Results..42

6 Summary...43

References/Bibliography..46

Acronyms..48

List of Figures

Figure 1: Excerpt from the SAT-EE...4

Figure 2: Excerpt from SAT-ETC..8

Figure 3: The Impact of Technology Adoption...33

Figure 4: Shewart Cycle..35

Figure 5: Technology Adoption Process...36

Executive Summary

To assist the Small Manufacturing Enterprise (SME) in adopting Advanced Engineering Environments (AEEs), this report provides two self-assessment tools.

The first tool, the Self Assessment Tool for Engineering Environments (SAT-EE) assists an SME in assessing the adequacy of the current computing support environment in handling technical tasks (i.e. Computer Aided Design (CAD) / Computer Aided Engineering (CAE) / Computer Aided Manufacturing (CAM)). This information can then be used to support decision-making with respect to expanding or upgrading the environment along the migration paths presented in this report.

The second tool is the Self Assessment Tool for Engineering Tool Capabilities (SAT-ETC). This tool collects the needs and desires of the operation, and maps them to the capabilities of specific classes of CAD and CAE tools, showing the SME the utility of these tools to his needs.

Many SMEs evolve or migrate from lower AEE levels to higher ones. As new tools and capabilities are introduced into a company's support environment on an incremental and opportunistic basis, the SME must recognize the shifts in operational modes and philosophies will be needed to gain the full benefit of the AEE.

In choosing a tool to be added to the SME's engineering environments, the SME must consider the functional capabilities of the tool, based on current and expected future needs; however, the SME must also take into account potential changes in business and software. The SME must also ensure that the tool's interoperability is appropriate to support the organization's overall product development process. Usability, particularly the extent to which the tool is deemed "intuitive" and reasonably "transparent" by its potential users, is a key factor in tool selection. Expandability, in terms of tool features that may be subsequently added and customization that users may apply to extend the tool's capabilities, should also be considered.

Selection, procurement, and installation of a new technology will not produce the desired benefits for the SME, until his staff adopts the technology and integrates it into its operation. Only when the SME's staff is aware of the technology, has access to it, is trained to use it, gets support for using it, and actually uses it will the benefits accrue to the SME.

A six-step process based upon the Shewart Cycle is presented. The process has been used successfully to introduce advanced technologies into SMEs and consists of the following steps:

1. Understand existing company environment;
2. Establish technology adoption project goals & metrics;
3. Evaluate technology options;
4. Obtain technology;
5. Implement & adopt technology; and
6. Analyze & deliver adoption results.

Abstract

To assist the Small Manufacturing Enterprise (SME) in adopting Advanced Engineering Environments (AEEs), this report provides two self-assessment tools; the Self Assessment Tool for Engineering Environments (SAT-EE) to assist an SME in assessing the adequacy of the current computing support environment in handling technical tasks, and the Self Assessment Tool for Engineering Tool Capabilities (SAT-ETC) to collect the needs and desires of the operation, and map them to the capabilities of specific classes of CAD and CAE tools. These tools help the SME evolve from lower AEE levels to higher ones, shifting operational modes and philosophies to gain the full benefit of the AEE.

In choosing an AEE component the SME must consider the tool's functional capabilities, interoperability, usability, and expandability. Selection, procurement, and installation of a new technology must be followed by tool adoption to integrate it into the SMEs operation. A six-step process based upon the Shewart Cycle has been used successfully to introduce advanced technologies into SMEs.

1 Introduction

1.1 Context of report

This report is the second in a series of two addressing Advanced Engineering Environments (AEEs) and their application to Small Manufacturing Enterprises (SMEs).

1.2 Purpose

The purpose of this two-volume report is to build awareness of the AEE concept, and assist SMEs in evaluating the desirability and feasibility of incorporating an AEE into their business operations.

Volume I of this report presented:

- candidate architectures for AEEs and comments on their applicability to SMEs;
- the benefits that may accrue to an SME from the adoption of an AEE in terms of internal and external effects;
- technical considerations that enter in the decision to adopt or upgrade an AEE;
- issues that an SME must consider when incorporating an AEE into his operation; and
- the general characteristics and capabilities of architectural elements or components for AEEs targeted upon geometry-centric design efforts.

Thes purpose of this volume is fourfold:

- to assist the SME in assessing his current status and future needs with respect to AEEs;
- to outline a series of migration steps whereby an SME may incrementally augment and expand the software environment that supports the organization's design activities;
- to define a method for selecting and adopting the Commercial Off-the-Shelf (COTS) tools that comprise an AEE; and
- to define a process to aid the SME in adopting a new technology.

2 Self-Assessment of AEE Requirements

This section presents two tools that may be used by an SME to assess its current status and future needs with respect to AEEs and their component computing tools:

1. Self Assessment Tool for Engineering Environments (SAT-EE) – this tool assists an SME in evaluating the adequacy of computing support for his current design operations; and
2. Self Assessment Tool for Engineering Tool Capabilities (SAT-ETC) – this tool assists an SME in understanding the value of various classes of engineering tools to its current and future design operations.

2.1 Self-Assessment Tool for Engineering Environments

2.1.1 Introduction

As discussed in Volume I, Chapter 2 of this report, an AEE may be categorized at one of three levels:

1. Basic   AEE;
2. Intermediate   AEE; and
3. Comprehensive   AEE.

The intended purpose of this section is to provide a tool whereby an SME could assess the adequacy of its computing support environment in handling its technical tasks (i.e., Computer Aided Design (CAD) / Computer Aided Engineering (CAE) / Computer Aided Manufacturing (CAM)) and could make decisions with respect to expanding or upgrading the environment along the migration paths presented in Chapter 3.

The initial concept for this section was a form of a predictive or normative tool that could make deductions of the form "If your company designs/produces X, then it should be running engineering support environment Y." On reflection, it became clear that dozens of aspects of the company's business would have to be considered to define "X" with sufficient precision and detail to be able to determine the various aspects of "Y" discussed in the previous reports with some degree of specificity. Furthermore, limited trials showed that even the nature of the aspects to consider is not clear. For example, one would expect that the nature of the product being manufactured may have some bearing on the outcome, but how would one characterize the differences in the required computational support between, say, a foundry, a welding shop, and a metal stamping plant? It was finally concluded that we don't have the knowledge and range of expertise to produce a credible predictive tool.

Attention thus turned to a less ambitious, but more doable descriptive tool which could be used to provide a rapid, albeit very coarse, assessment of the company's current practices with respect to the environment supporting design and engineering, and of the adequacy of the currently available engineering environment. The remainder of this section deals with the resulting self-assessment tool.

See the SEI website (http://www.sei.cmu.edu/tide/publications/SAT-EE.XLS) for an interactive version of this tool. An excerpt from the tool is seen in Figure 1.

SELF-ASSESSMENT TOOL FOR ENGINEERING ENVIRONMENTS (SAT-EE) v0.6

© 2003 by Carnegie Mellon University

RESULTS

COMPOSITE SCORE		2.0
APPROXIMATE LEVEL OF ENVIRONMENT		intermediate level
IMPROVEMENT OPPORTUNITIES		none identified

Category	Units	Response	Score
CAD			**2.2**
CAD tool type			5.0
Do you use a CAD tool?	y/n	y	
Desktop CAD	y/n	y	
Light-duty CAD	y/n	y	
Heavy-duty CAD	y/n	y	
CAD add-ons			1.0
Purchased or rented symbol and/or detail libraries	y/n	n	
Purchased or rented discipline-specific add-ons	y/n	n	
Internally developed libraries and/or add-ons	y/n	y	
CAD usage			3.6
% of designers that have CAD seats	%	80	
% of design information received from suppliers as CAD files	%	20	
% of design information sent to customers as CAD files	%	30	
% of design information sent to manufacturing as CAD files	%	10	
% of design information stored as CAD files reused on new projects	%	30	
CAE			**1.7**
CAE tool type			2.5
Do you use a CAE tool? (as opposed to spreadsheets, standalone programs, analysis consultants, etc.)	y/n	y	
Light-duty CAE	y/n	y	
Heavy-duty CAE	y/n	y	
Does the CAE tool include special features for particular kinds of applications (e. g., sheet metal forming, non-linear analysis, etc.)?	y/n	y	
Does the CAE tool support multiple analysis disciplines (e.g. stress, thermal, electromagnetic, etc.)?	y/n	n	
CAE usage			3.3
% of engineers that have CAE seats	%	20	
% of designers that have training in and use CAE	%	0	
% of analyses performed by outside consultants	%	20	
% of design information shared with suppliers and customers in the form of CAE models and results (as opposed to information shared in the form of CAD models, drawings, etc.)	%	30	
Is CAE used in proposals and/or preliminary design?	y/n	y	
Is CAE used in detailed design?	y/n	y	
What is the average number of design iterations based on feedback from CAE analyses?	number	3	

Figure 1: Excerpt from the SAT-EE

2.1.2 Function of the tool

The tool collects information about the engineering environment at the SME in nine categories:

1. CAD (Computer Aided Design)
2. CAE (Computer Aided Engineering)
3. CAM (Computer Aided Manufacturing)
4. CAD/CAE integration[1]
5. CAD/CAM integration
6. PDM (Product Data Management)
7. CAD/PDM integration
8. Catalog access
9. Database support

In each category, information is collected about two sub-categories:

   a. the type of tool used in that category; and
   b. some characteristic statistics about the extent of the tool's usage. In the case of CAD, there is a third sub-category of information collected about the add-ons to the off-the-shelf CAD systems.

Based on the information entered, the tool assigns:

- a composite numerical score, in the range of 0 to 5, for each of the nine categories;
- a weighted aggregate score, again in the range of 0 to 5, for the environment as a whole;
- an approximate determination of the level of the environment (below minimum, minimum, intermediate, above intermediate); and
- some recommended changes in the environment.

It is to be emphasized that the tool is very empirical and approximate in nature:

- the information collected is by no means exhaustive for the intended purpose and may not even be appropriate for the intended evaluation;
- the weights and weighing functions assigned to the individual responses may not be appropriate;
- the aggregation function for the composite score may not be appropriate; and
- the function for assigning environment levels is highly simplistic.

---

[1] Readers of the first report in this series will recognize that the tool is geared towards the Basic AEE summarized in Section 2.1.1 of that volume: it is assumed that the CAD system, rather than the database system, acts as the central repository of product information.

Nevertheless, the tool is presented as a first cut at the problem. A future usability study may be used to evaluate its effectiveness and refine its categories, questions within the categories, weights, and weighing functions.

2.1.3 Use of the tool

1. To use the tool, activate it as an Excel spreadsheet. The tool is compatible with Microsoft→ Excel 2002. Select the "Assessment Tool" worksheet of the tool.

2. All areas requiring user input are color-coded in bright yellow. Initially, all the responses (column F) are blank and all sub-category and category scores display the "error" message.

3. ENTER A RESPONSE TO EVERY QUESTION (within reason - if you answered "no" to the question "Do you use a CAE tool?", you need not answer the remaining questions in the sub-category pertaining to the CAE tool – however, in other categories there may be multiple responses – a company may be using desktop, light and heavy-duty CAD systems simultaneously).

4. Fields marked "y/n" in the "Units" column (color coded in pale yellow) respond only to a "y" or an "n"; those marked "%" respond to an integer between 0 and 100; those marked "number" respond to an integer value.

5. When responses for a category are completed, the composite scores for the category and its sub-categories are displayed. When all categories are completed, the aggregate score, the approximate level of the environment, and the recommended changes in the environment are displayed.

6. You may experiment with alternate scenarios by clearing some or all of the responses and entering new values:

    a. to clear all responses: click on the pull-down menu of the Excel *Name Box* (located near the upper left corner of the spreadsheet on the left side of the "*Formula Bar*"); click on "AllResponses"; and press "Delete".

    b. to clear values in a particular category: click on the pull-down menu of the *Name Box*; click on the name of the appropriate response category (e. g., "PDMResposes"); and press "Delete".

2.1.4 Interpretation of Results

The Composite Score (cell H6) presented by the tool is a numerical approximation of the current AEE level of the respondent. AEE levels are derived from this score as follows:

$$\text{Composite Score} < 0.9 \Rightarrow \text{below Basic AEE level}$$
$$0.9 \leq \text{Composite Score} < 1.8 \Rightarrow \text{Basic AEE level}$$
$$1.8 \leq \text{Composite Score} < 3.4 \Rightarrow \text{Intermediate AEE level}$$
$$3.4 \leq \text{Composite Score} \Rightarrow \text{Comprehensive AEE level}$$

→Microsoft and Excel 2002 are registered trademarks of Microsoft Corp., Redmond, WA

The Composite Score is calculated as the weighted average of the 9 category scores. Weighting is as follows:

Category	Weight
CAD	2
CAE	1
CAM	1
PDM	1
CAD/CAE Integration	1
CAD/CAM Integration	1
CAD/PDM Integration	1
Catalog Access	1
Database	1

Examination of the category scores will reveal areas for potential improvement that could enhance the AEE level of the organization. The user may experiment with these areas to assess the impact of proposed changes. The SME must evaluate these potential improvement areas in the context of his business strategy to achieve meaningful operational improvement.

2.2 Self Assessment Tool for Engineering Tool Capabilities

2.2.1 Introduction

The intended purpose of this section is to provide a tool whereby an SME can assess its needs for computational tools for technical (CAD/CAE) tasks and can assist in identifying the additional needed tools according to the criteria discussed in Chapter 4.

The initial concept for this section was for a multi-level tool that would:

1. Identify SME needs that could be addressed with an Advanced Engineering Environment (AEE) or one of its components;
2. Identify SME problems that could be addressed with an AEE;
3. Map SME problems and needs to generic AEE functions; and
4. Map generic AEE functions to classes of tools comprising AEEs.

The first two steps turned out to produce extremely long lists of hypothetical needs and problems without much structure. As soon as some structuring was attempted, it became clear the next two steps needed to be collapsed: AEEs have very little functional redundancy, and a need can be addressed or a problem solved by only one software component of the AEE. The capabilities of the AEE component tools provide the organizing principle for the needs and problems.

Thus, the needs assessment tool became less ambitious, but more practicable as a descriptive tool which could be used to provide a rapid, albeit very coarse, assessment of the company's needs for – and desires – for problem-solving capabilities contained in AEE component tools. The remainder of this section deals with the resulting self-assessment tool.

See the SEI website (http://www.sei.cmu.edu/tide/publications/SAT-ETC.XLS) for an interactive version of this tool. An excerpt from the tool is seen in Figure 2.

Category	Response						Score	
	Very valuable in routine use today	Valuable in occasional use today	Marginally valuable capability available but we have no training	Potentially very valuable would use immediately if we had access to capability	Potentially valuable would consider using if we had access to capability	Not valuable or unable to comment	Current value	Potential value
MECHANICAL CAD							2.50	3.01
3D Component Design							1.7	2.1
A1 Create 3D models of mechanical and structural components	y							
A2 Define and use "parametric" 3D component models constructed such that a change in a dimensional "parameter" results in a change of the design and appearance of the component (i.e. changing the dimension of a hole diameter automatically resizes the hole shown in the design)		y						
A3 Define and use constraints on 3D component models (e.g. specify that two surfaces are parallel, specify that two cylinders are concentric, etc.)				y				
A4 Compose component models using saved features, catalog entries, and other components			y					
A5 Assign/display component dimensions and tolerances	y							
A6 Assign/display component annotations such as dimensions, tolerances, explanatory notes Bill-of-materials reference numbers, etc.					y			
A7 Assign/display non-geometric attributes of components such as material properties, surface finishes, etc.						y		
3D Assembly Design							0.0	4.2
B1 Generate 3D sub-assemblies and assemblies by merging component models				y				
B2 Define interfaces between components parametrically. For example, two flanges may be connected by a circle of "N" bolts. The value of N defines the number and spacing of the holes in both components, as well as the number of bolts on the BOM.					y			

SELF-ASSESSMENT TOOL FOR ENGINEERING TOOL CAPABILITIES
© 2003 by Carnegie Mellon University

Figure 2: Excerpt from SAT-ETC

2.2.2 Function of the tool

The tool collects information about engineering problem-solving needs in the following categories and sub-categories:

1. MECHANICAL CAD
    a. 3D Component Design
    b. 3D Assembly Design
    c. 3D Model Reuse
    d. 3D Model Post-Processing – Graphical
    e. 3D Model Post-Processing – Non-Graphical

2. CAE – FINITE ELEMENT ANALYSIS
    a. Analysis Model Generation and Import
    b. Analysis Model Idealization and Discretization
    c. Analysis Model Reuse
    d. Analysis Options
    e. Analysis Purpose
    f. Structural Analysis Options
    g. Computational Fluid Mechanics Analysis Options
    h. Electrostatic/Electromagnetic Analysis Options
    i. Thermal Analysis Options
    j. Multiphysics Analysis Options
    k. Post-Processing of Analysis Results

3. CAD-CAE INTEGRATION
    a. Importing from CAD to CAE
    b. Idealization of CAD model into CAE Model
    c. Post-Processing of CAE Model
    d. Exporting from CAE to CAD

In each sub-category, a number of problem-solving capabilities are listed and the user is asked to rate each capability as one of:

- Very valuable, in routine use today;
- Valuable, in occasional use today;
- Marginally valuable, capability available but we have no training;
- Potentially very valuable, would use immediately if we had access to capability;
- Potentially valuable, would consider using if we had access to capability; or
- Not valuable or unable to comment.

Based on the information entered, the tool assigns a Current Value and a Potential Value, on a scale of 10.0 to 0.0, to each sub-category, then to each of the three main categories, and finally to the response as a whole.

2.2.3 Use of the tool

1. To use the tool, activate it as an Excel spreadsheet. The tool is compatible with Microsoft→ Excel 2002. Select the "Assessment Tool" sheet of the Workbook

2. Initially, all the responses (columns C through H) are blank and all sub-category and category scores display the "error" message.

3. Enter your responses. Enter only one "y" in each row. A character other than "y" does not contribute to the score.

4. When responses for a category are completed, the composite scores for current and potential values of the category are displayed.

5. When all categories are completed, the aggregate scores are displayed.

6. You may experiment with alternate scenarios by clearing some or all of the responses and entering new responses:

    a. To clear all responses: click on the pull-down menu next to the Name Box (leftmost box in bottom row of Excel banner - normally displays the designation of the currently active cell); click on "All Responses"; and click on "Delete".

    b. To clear responses in a particular category: click on the pull-down menu next to the Name Box; click on the name of the appropriate response category (e. g., "CAD Responses"); and click on "Delete".

2.2.4 Interpretation of Results

Based on the information entered, the tool assigns a Current Value and a Potential Value, on a scale of 10.0 to 0.0, to each sub-category, then to each of the three main categories, and finally to the response as a whole. The aggregated scores are not as meaningful as in the self-assessment tool presented in the previous section. The user is advised to scan the sub-categories with high scores for Potential Value – these are the tools and tool capabilities that should influence the tool selection process described in Chapter 4.

It is to be emphasized that the tool is very empirical and approximate in nature:

- The problem-solving capabilities in each sub-category are not exhaustive;
- The set of sub-categories for each main category may not be exhaustive;
- The set of main categories was purposely restricted to three – mechanical CAD, CAE using Finite Element Analysis (FEA), and the integration of the two – in order to make the needs self-assessment tool manageable in size; and
- The user is asked to treat sub-categories independently of each other, when in reality many of the sub-categories are coupled – in particular, Analysis Purpose and Options are

strongly coupled with the domain-specific options as well as the Integration sub-categories.

Nevertheless, the needs self-assessment tool is presented as a first cut. A future usability study may be used to evaluate its effectiveness and refine its categories, sub-categories, problem-solving capabilities, rating scale and rating aggregation.

3 Migration

The purpose of this chapter is to outline a series of migration steps whereby an SME can incrementally augment and expand the software environment that supports the organization's design activities as well as its linkages to other corporate activities.

This chapter describes the migration of an organization's computing environment in distinct steps, each step involving one transformation of the environment, rather than in terms of longer paths that may include several transformations. Each of the steps described requires considerable learning, experience- and confidence-building on the part of all the participants before the organization is ready to take the next step. It is recommended that a period of approximately two to three years be spent between consecutive steps, allowing for at least two full product development cycles to take place. Although the steps are described in this report as discrete ones, many organizations expand their computing environments incrementally, as new needs or corporate functions arise and as new capabilities offered by software vendors become technically and economically justifiable. Nevertheless, the environments defined by the steps outlined in this report represent fundamentally different modes of operation and philosophical outlooks on computer-aided product development. The environments resulting from the steps described in detail below are considered to be prototypical of the development of engineering environments over the past two decades.

Each step is described by means of a standard template, consisting of the following components:

- *Synopsis of Situation*: a brief description of the characteristic features of the organization's current computing environment;
- *Symptom*s: signals, both internal and external, that indicate that the current situation is becoming difficult or uneconomical to maintain and that a transformation to a higher level is warranted;
- *Alternatives*: aspects of the transformation that need to be considered, and a partial list of the alternatives available for selection in each aspect;
- *Decisions/evaluations*: the criteria to be used for each selection and the technical considerations that may enter into the ranking or selection of the alternatives;
- *Training needs and other preparations*: the technical, organizational and personnel issues to be addressed in preparation for the migration; and
- *Results to be expected*: a brief description of the changes that may realistically be expected to occur after the new environment is put in place.

3.1 From scratch to CAD for drafting

Synopsis of Situation: This situation is rare today, but may exist in some small manufacturing enterprises. Computers may be used in accounting, parts inventories, etc., and partial CAD modeling may be practiced for Numerically Controlled (NC) machining purposes. No CAD tools are used for design, all design documentation being in paper-based drawings. Engineering design functions may use spreadsheets or some stand-alone tools.

Symptoms: Maintenance of paper drawings is cumbersome; it is hard to search drawings for design reuse; complete redrawing is necessary even for the most minute design modification. Lack of analyses necessitates repeated physical prototype construction and testing. Paper-based communication is inefficient, both vertically (e. g., with clients, the manufacturing division, suppliers) and horizontally among design and engineering groups.

Alternatives: Clearly, the prime issue to confront is the installation of a CAD system. The options to be explored are:

- CAD system level, which may be:
    - Entry level, sometimes referred to as desktop;
    - "Light" version of one of the major CAD systems; or
    - Full strength version.
- The extent of CAD system customization that will be needed. The options are:
    - No customization, use the CAD system "out of the box";
    - Purchase or rent symbol and/or detail libraries from CAD vendor or third party;
    - Purchase or rent discipline-specific add-ons (e.g., sheet metal drafting) from the same sources; or
    - Contract out for the development of custom libraries.
- The extent of CAD/CAM integration to be achieved. The options are:
    - Install same tool or platform for CAD and CAM, with different add-ons;
    - Plan on using different tools for CAD and CAM and interfacing (exchanging models between) them (e. g., via STEP); or
    - Initially provide no interfacing.
- Personnel allocation. This may be an even harder choice than software/platform selection. The extreme points of alternatives are:
    - No segregation of duties: all designers will operate CAD system on their own; or
    - Separation of functions between designers and CAD station operators.

Decisions/evaluations:

- The three choices of CAD level increase both in cost and operator training requirements. Entry level systems provide the least demanding transition from paper-based design, in both system cost and training, but their performance can easily degrade when working on larger CAD files typical of commercial products. Mid-range versions, while somewhat more demanding in training, are intended to offer a smooth transition to the next level, should that be warranted.

- The choices in customization also increase in cost, but they also significantly increase the efficiency, productivity and satisfaction of the users. Purchase, rental or acquisition as freeware of symbol and detail libraries, and generally of add-ons, provides big dividends in productivity at small incremental cost. Large amounts of customization are to be avoided. At this stage of low in-house familiarity with the tools, the organization can become overly dependent on the provider of the customized software.

- Whether the products designed are manufactured within the organization or by outside suppliers, CAD/CAM integration should be adopted from the outset. Unless there is some legacy software that warrants interfacing between CAD and CAM tools, it makes sense to adopt the same platform from the start.

- Personnel allocation choices should reflect the enterprise's policies and working conditions. In an informal, task- or project-oriented design department, it makes sense to have all designers access the CAD system directly, whereas in a more hierarchically structured workplace specialized CAD station operators, who are trained to be experts in using the tools, would make more sense.

Training needs and other preparations: The most important preparation has to do with personnel: selection of the personnel allocation method; training and/or hiring of personnel; and planning and training for the specialized functions that will arise (e. g., CAD system manager, CAD system maintainer, CAD file archivist, etc.). The introduction of a new mode of doing business warrants, even mandates, a thorough review of the organization's processes, and if necessary, a reorganization. Finally, a plan needs to be developed and maintained for the upgrading and expansion of the engineering environment and the orderly replacement of its components.

Results to be expected: After some initial training and a learning curve, the organization will find that it takes less time to generate even initial drawings. Search for and modification of drawings will be drastically improved. Electronic storage of drawings will be found to be convenient and will rapidly lead to increased design reuse. Generation of derived information, such as assembly drawings and BOM, will be easier and much more error-proof. Eventually, the advantages of a smooth transition to downstream processes, especially with 3D models (CAE, CAM), will become obvious. Communication with other designers, engineers and clients by means of CAD data will be easier.

3.2 From CAD for drafting to CAD and external analyses

Synopsis of Situation: 2D and/or 3D CAD tools are used for drafting and for the representation and communication of the product's geometry only. In all other respects, conventional or "traditional" engineering practice prevails, except for the substitution of CAD tools for drafting boards.

Symptoms: Rough analyses are carried out using design manuals, tables, formulas and possibly spreadsheets and stand-alone programs. Whatever analysis model there is, it is manually built by the engineers, with extensive dependence on the engineers' expert knowledge. Analysis is even harder for innovative designs or new design configurations about which there is less expertise. The products tend to be over-designed due to the engineers' lack of confidence in predicting the product's performance, potentially increasing product cost. Designs are verified with costly physical prototypes. There is slow response to customers' Requests for Quotation (RFQs) that have substantial technical specifications. Design reuse is not easy due to the separate management of design and analysis data.

Alternatives: The prime issue here is the improvement of CAE analysis capability commensurate with the improved geometry manipulation capability afforded by a CAD system. The two major alternatives are to engage an external consultant or to jump a step in the progression discussed in this report and initiate in-house analysis competence. The remainder of the discussion addresses the first alternative; the second one will be presented in the discussion of the succeeding step. The major issues in selecting an external consultant are:

- Consultancy provider type:
    - National firm;
    - Local firm or branch;
    - Small firm or individual; or
    - Application Service Provider (ASP).
- Timing of analyses in the design cycle:
    - Early, e.g., in preliminary design or even response to RFQ;
    - Late, typically after detailed design is completed: or
    - Several stages of the design process.
- Scope of consultant functions:
    - "Turnkey" operation: consultant models, idealizes, analyzes, interprets model and recommends any design changes needed;
    - Consultant given a CAD model and only idealizes, analyzes and interprets results; or
    - Modeling, idealization and interpretation are done in-house, consultant only performs the analyses.
- If multiple functional domains are involved (e. g., structural, thermal, fluid flow, etc.), does the organization engage:
    - One analyst for all domains; or
    - Several analysts, one for each domain.
- Disposition of existing in-house analysis software;
    - Continue its use;
    - Cancel further use; or
    - Develop in-house software for checks on consultant's results.

Decisions/evaluations:

- Concerning the prime issue of in-house vs. external analysis, external capability is considered to be better where the current in-house analysis activity is minimal and the size of the enterprise is not large enough for maintaining analysts and analysis tools.

- Concerning provider type, the main considerations are ease of access (physical or virtual), consultant availability when needed, and continuity of expertise. Small consulting firms tend to have higher staff turnover than large ones, and individuals, whether "moonlighting" or not, tend to be more tied up with on-going tasks than larger consulting firms. ASPs are a new phenomenon: they are certainly good candidates for providing "raw" analysis capabilities, but their performance at other levels (e. g., idealization and interpretation) has not yet been well established.

- While it may be highly desirable to involve consultants early in the design process, the normal turnaround time, particularly in the "turnkey" mode, makes this unsuitable, if not impossible. By far the most typical use of analyses performed by consultants is for design verification after the detailed design is (essentially) complete. With proper planning and coordination the consultant may be brought in early, given time to develop a model "template," and then the template may be used repeatedly as the design process unfolds.

- The alternative scopes range from essentially full involvement by the consultant to essentially full in-house involvement, "farming out" only the resource-consuming "number crunching" part of the analysis. The choice will be governed by the extent of internal experience and expertise; typically, as this expertise develops, the dependence on outside consultants decreases.

- If multiple domains are involved, it is probably easier to manage interactions with one consultant than with several ones; however, there may be cases where the specialized expertise needed in some of the domains may only be available from firms specializing in that domain only.

- Concerning in-house software, it certainly does not make sense to perform in-house analyses on tasks that a consultant has been engaged on (more appropriately, it does not make sense to engage an external consultant for tasks for which there is available in-house expertise supported by suitable tools). Typically, however, in-house tools will continue to be used in the early design stages where there is not yet a detailed enough model to turn over to the consultant, and it is imperative that tools be developed or acquired for performing coarse or approximate checks on the consultant's detailed results.

Training needs and other preparations: In terms of personnel needs, even for the turnkey mode, a few engineers knowledgeable about the analysis task will be needed; introductory seminars or short courses are advisable for all engineering personnel. Training must be provided to engineers so that they can evaluate analysis models, interpret analysis results, and accept analysis feedback to modify designs (if production time permits). Training also needs to be provided for CAD modelers so that they produce "good" geometric models conforming to CAE tool needs. The engineering process will have to be reviewed, and if necessary adjusted, to accommodate the consultant's turnaround time, which in the "turnkey" mode may be of the order of weeks. In addition to purely process changes, management will have to address issues of professional responsibility and establish a responsibility chain for the technical performance of the products. Plans for future migration to internal analyses should be in place early in the transition.

Results to be expected: More detailed and thorough analyses with up-to-date analysis tools will provide increased confidence in the performance of the company's products. Design verification will be less costly and time-consuming with the possible elimination of some physical prototypes. Templates of prototypical products will provide faster turnaround on repetitive product types. On the other hand, analyses for innovative designs or new product configurations will be much easier with proper analysis tools. Design reuse will be easier with design and analysis data managed more closely. The organization will find that it can automate substantial portions of the analysis process, becoming less dependent on the engineering experts' knowledge.

3.3 From CAD and external analyses to a Basic AEE

Synopsis of Situation: Engineering analysis is dependent on external contractors, even though in-house engineers have gained increased knowledge of CAE processes, particularly on how to interpret analysis results and feed them back for design enhancement. The organization recognizes the importance of engineering analyses throughout the design process and is ready to invest in CAE tools and personnel.

Symptoms: The number of design/analysis iterations is limited due to the long turnaround cycles. Design optimization, whether for performance or cost, is hard to achieve due to inefficient communication between designers and external analysts. Engineering knowledge is not accumulated systematically for future reuse. There are concerns about the security of proprietary intellectual property communicated to the external consultants.

Alternatives:

- The prime issue is the "internalization" of the analysis capability by the installation of a CAE system. The options to be explored are discussed extensively in Report 3 in terms of the following aspects:
    - Choice of depth in terms of levels of the design process (e. g., conceptual, preliminary, detailed) to be supported;
    - Choice of breadth in terms of the number and kind design sub-disciplines to be supported;
    - Choice of CAE components of the environment;
    - Choice of component specialization (sheet metal, injection molding, etc);
    - Choice of COTS products; and
    - The degree of customization that will be required.
- Personnel allocation. As in the case of the CAD system selection, this may be an even harder choice than the CAE component and COTS product selection. Again, the extreme points of alternatives are:
    - No segregation of duties: all engineers use the CAE system(s) on their own; or
    - Separation of functions between engineers and CAE analysts.

Decisions/evaluations:

- For a discussion of criteria and options on CAE tool selection, see Report 3.
- Personnel allocation choices should again reflect the enterprise's policies and working conditions. In an informal, task- or project-oriented engineering department, it makes sense to have all engineers use the CAE tools directly, with only a very few full-time analysts, whereas in a more hierarchically structured or compartmentalized workplace, dependence on a dedicated analyst group would make more sense. In either organizational mode, interaction between engineers and analysts can be enhanced by providing "light" versions of the CAE tool to the first group and "full-strength" versions of the same tool to the second group.

Training needs and other preparations: Personnel decisions will include: selection of allocation method: training and/or hiring of personnel; decisions on specialization of functions between analysts and designers. Engineers need to be trained to build and use coarse-grain analysis models suitable for proposals and conceptual designs. Engineers also need to be trained to use feedback from analysis results to modify designs iteratively and to do sensitivity analyses. The design process will have to be reviewed and/or reorganized to move analysis upstream into the early design stages and to organize the design process for design iterations. The organization must develop policies for determining when a design is considered "good enough" without further iterations. As always, the firm has to plan for replacement, upgrades, and expansion of the environment.

Results to be expected: The organization will find that it can set up efficient engineering processes with tools suited for its specific need. It will be able to provide faster responses to customers' RFQs with demanding technical specifications, thus gaining a competitive edge in the marketplace. Design and analysis processes will be coupled, providing for easier preparation of analyses and rapid turnaround. It will become easier to optimize designs for performance and cost, and to reuse engineering knowledge.

3.4 From Basic AEE to Intermediate AEE

Synopsis of Situation: Design is driven primarily by geometry: CAD models and drawings are at the center of the product data representation, with analysis processes weakly linked to the spatial design. Design rationale is not captured in the product data representation. There is no systematic creation and management of a complete and persistent representation of the evolving product model from the earliest conceptual design steps to the completion of the detailed design and beyond to manufacturing.

Symptoms: Design intent and knowledge applied are not captured in the product representation. Design reuse is difficult without access to the design intent embedded in the product data. Lack of a central product data model, encompassing both geometry and engineering function/behavior information, makes sharing information between different disciplines inefficient and collaboration hard to achieve. With the predominance of geometry, most engineering analysis activities can start only after the CAD model is fairly well developed; engineering activities are not easily accommodated during the early stages of the design evolution process.

Alternatives: The prime issue is the achievement of seamless two-way interoperation among all tools, the CAD tool included. This can de achieved in three ways: direct tool interfacing, interfacing through translators or interfacing via a database. The alternatives to explore are:

- Direct tool interfacing through a common "native" language;
- Interfacing through translators and/or neutral files e.g., STEP (Standard for the Exchange of Product model data), IGES (Intermediate Graphic Exchange Standard); or
- Interfacing via a database system; in this case, further decisions need to be made on the scope, location and nature of the database (highly unlikely that the development of such a database system will be a viable choice for an SME).

Decisions/evaluations:

- In the direct interfacing option, at today's state of technology, direct interfacing through a common "native" language is possible only if all tools are provided by the same vendor or consortium of vendors, which is rarely the case for a realistic array of CAD/CAE/CAM tools.
- The alternate interfacing options fall into two classes: tool-to-tool two-way direct translators and translators to and from a central representation often called "neutral files." A true "neutral file" capable of interfacing with all tools used in a firm would, in many respects, be functionally equivalent to the shared database of the second alternative.
- In the database option, a host of issues would have to be addressed. Today, only large corporations can dedicate the resources needed for building custom databases, and even they have largely switched to buying COTS systems that are "tailored" by the vendor, or a third party, for them. An SME would typically be buying one such system.

Results to be expected: The organization will be able to create and manage persistent representations of the evolving product model from start to finish. There will be efficient support of all engineering analysis activities during the entire design phase. Design reuse will be dramatically improved by the shared storage of geometry and engineering design data. The link between design and analysis processes will be tightened to the point that integrated design-analysis, function-driven design and multifunctional design-analysis can be routinely performed.

3.5 Beyond an Intermediate AEE

Synopsis of Situation: This is largely uncharted territory, because only a few of the largest manufacturing enterprises (e. g., in the automobile, aerospace, and defense industries) have reached this level, and extrapolation to SMEs is difficult. Organizations at this level maintain

CAD independent representation of the product model as a core, often augmented with the representation of design rationale. Application-specific data may also be maintained in parallel for tighter integration, mainly for analysis applications. Extensions such as PDM and catalog facilities are also generally used.

Symptoms: Product data are effectively utilized only inside the design and analysis departments. The environment does not support a global engineering architecture. There is a routine need for integrating internal and external (contracted out) designs, and for configuration management of a complex product or product suite over many engineering units, both internal and external. There are serious problems in communication between the core activities integrated into the environment and various vertical and horizontal applications external to the environment.

Alternatives: The alternatives involve either incorporating new tools and/or new applications into the current environment, thereby expanding the environment through new components. Function-driven design tools, knowledge-based CAD systems, multidisciplinary simulation and synthesis systems are potential examples in the first category. Immersive CAD technology, virtual manufacturing and collaboration support technologies are examples of AEE components that may be added.

Decisions/evaluations: There is no established precedent for selecting among the alternative expansions or their constituents.

Training needs and other preparations: Again, there is no established precedent. There are training needs for the new skills introduced by the new components, and training needs for existing personnel to integrate the new tools in their tasks.

Results to be expected: Better management of product development processes, including the creation, monitoring and modification of design documents and databases, may be expected. More thorough evaluation of the operability, manufacturability and maintainability of the proposed designs may be performed as part of the design process. Seamless sharing of product data in a distributed and heterogeneous engineering environment, and effective collaboration through various communication channels between agents and repositories will become the mode of operation.

3.6 Additional migration considerations

Two additional migration considerations, namely, downstream data integration and PDM adoption, are presented separately because they are essentially independent of the level of the engineering environment. The two issues may be addressed at any of the levels discussed, either in conjunction with one of the transformations discussed above or entirely separate from them.

3.6.1 Downstream (CAD/CAM) integration

Synopsis of Situation: The organization does not use any CAM tools, or separate CAD and CAM tools are used without careful consideration of their integration or interoperation.

Symptoms: There are large amounts of rework in the CAD system or manual entry to the CAM system because of the presence of incompatible geometric models, typically encountered late in the product delivery process when NC code generation for the parts is first attempted. Due to the lack of linkage to the CAD system, machining operations suboptimal in quality, time and/or cost may result.

Alternatives: The issue is a simple one: establish CAD/CAM integration/interoperation. The alternative approaches are:

- Shared platform for both CAD and CAM;
- Interfaced platforms with vendor-supplied built-in interfaces;
- Shared files (e. g., STEP); or
- Integration via PDM (see below).

Decisions/evaluations: The choice among the alternatives listed will largely depend on external, non-technical, considerations such as:

- The nature of engineering/manufacturing interaction. If design and manufacturing are tightly integrated (in the extreme, no designs are manufactured outside, no manufacturing of outside designs) a shared platform supplied and maintained by a single vendor makes more sense. On the other extreme, if most of the designs are "farmed out" for outside manufacturing and/or manufacturing produces mostly designs of outside organizations, shared files in standard formats such as STEP may be the most practical.

- Familiarity with tools at both sites is another consideration. If either engineering or manufacturing has long been using one tool, it makes sense to install a shared or interfaced tool for both sites.

- The position in the supply chain. Sometimes the choice is made for the SME by a larger client dictating the CAD or CAM tool to its suppliers, or a large supplier making it advantageous to the client to have compatible tools. This factor may be less important in the future as web viewing tools will increasingly provide translations as well.

Training needs and other preparations: For engineering, the primary personnel training/hiring issue is that of training CAD modelers to produce "good" design models conforming to CAM tool and production needs. Quality control and change control policies and their implementations will have to be developed jointly between engineering and manufacturing.

Results to be expected: The standard CAD modeling practice previously used for design only will now also result in near-flawless NC code generation. Where CAM modeling detects that the CAD models are incomplete, CAD data healing technologies can be utilized.

3.6.2 PDM adoption

Synopsis of Situation: The organization uses CAD (and possibly CAM) tools, but product data, documents and drawings are dispersed over several systems and inefficiently managed. Design process management is treated entirely separately from the management of the data generated and used in that same process.

Symptoms: It is difficult to share design documents and product data among engineers working on different stages or disciplines in the design process. There may be duplicate data inputs and/or inconsistencies among different versions of the emerging product's design. Separate design process management and product data management either produce conflicts and contradictions, or are de-emphasized so as not to produce conflicts. Design reuse is hampered by the separation of process and product data.

Alternatives: The issue again is a simple one: introduce a PDM system that integrates design process management and product data management. The alternatives to consider are:

- Scope of PDM system:
    - Engineering only;
    - Enterprise-wide; or
    - Integrated with clients and/or suppliers.
- Integration into environment:
    - Stand-alone application (i. e., PDM data are entered and used separate from engineering and manufacturing applications); or
    - Integrated with CAD/CAE/CAM tools.

Decisions/evaluations: As with the CAD/CAM integration discussed above, both the choices of scope and the manner of integration with the design environment will largely depend on external factors, primarily on the nature of integration between engineering and management and on the organization's position in the supply chain (see discussion above). A separate PDM system makes little sense today when so many integrated systems are available. Access, via the web, to the client's or supplier's PDM system makes sense, but external access to the SME's engineering information needs to be evaluated and then very carefully controlled when such interfaces are provided, because PDM systems provide direct access and usually only monitor transactions by logging (i. e., after the fact).

Training needs and other preparations: Personnel hiring/training decision involve the personnel who will be hired to run the PDM system or who will be retrained for PDM from the organization's current process management functions. The process control policies and their implementation must be thoroughly planned to obtain the maximum value from the PDM integration.

Results to be expected: The organization can expect: increased productivity; improved product quality from fewer errors in product data and the potential of detecting downstream quality problems early in the design process; increased data security: rapid availability of information irrespective of product development stage or user location; and ease of search for parts and documents. These benefits will accrue from better version control and document tracking provided by PDM systems as well as better engineering process and change management introduced by the organization as part of the implementation of PDM.

3.7 Summary

The evolution of an organization's computing support environment was presented as a series of transformations between levels. Even if new tools are introduced into a company's support environment on an incremental and opportunistic basis, the levels first introduced in Report 1 and referred to throughout this series of reports represent fundamentally different modes of operation and philosophical outlooks on computer-aided product development. The question of what level of the computing environment deserves to be called an Advanced Engineering Environment is moot. A company that progresses two levels in the hierarchy presented here in a few years will view its environment as advanced.

4 Tool Selection

As discussed in Chapter 3, the largest step change in migrating from one level of AEE to the next is the acquisition of a new tool and the insertion of that tool into the SME's enterprise and design process. The purpose of this chapter is to outline a series of considerations that enter into the selection and acquisition of a tool to be added to an SME's engineering environment. The chapter presents only generic considerations that hold for all the tools discussed in this report, and does not deal with considerations of the specific selection criteria of classes of tools (e.g., criteria on the types of nonlinear analysis for selecting a CAE tool).

It is assumed that an evaluation of the SME's needs has confirmed the technical basis for acquiring a tool within a particular class, and that at least a preliminary analysis has confirmed the economic viability of that acquisition.

Tool selection is a multi-criteria decision-making problem, and almost every decision maker can use some assistance in the process. Therefore, the chapter contains a brief discussion of some of the resources on which an SME can draw.

Finally, there is the issue of tool granularity. The selection criteria presented here tend to assume a coarse-grained selection and acquisition process, e.g., selecting a CAD tool or a PDM tool, in a process similar to selecting a drafting table or a file cabinet. Today's AEE component tools are closely interfaced and bundled, and the choices to be made tend to be more fine-grained. The technical capability that one wishes to acquire may be available as an upgrade of or an add-on to one of the tools in the SME's current "toolbox." This fact may result in having to make much more heterogeneous choices, comparing add-ons for existing tools to brand new tools.

4.1 Selection criteria

4.1.1 Technical criteria

Functionality. Clearly, the first criterion for selecting a tool, or evaluating a potential tool, is that the tool provides the functionality that is needed by the organization. The needs assessment tool discussed in Chapter 2 is designed specifically to elicit some of the potential needs of an SME, and should be consulted first. However, the tool can obviously never be complete, and the SME will generally identify additional functionalities needed beyond those addressed by the tool. Reviews of past designs and of their documentation and discussions with designers and engineers are some of the means for identifying needed functionalities in-house. Tabular comparisons of tool features and vendor literature and presentations are the key means for establishing whether the tools offered provide the functionality needed.

In defining the needed functionality, it is important to consider an appropriate time frame for the tool being considered. Obviously, for investing in the purchase of a tool and the larger expense of providing adequate staff training in the use of the tool, the SME wants more functionality than just that needed to attack today's problems. It is necessary to do some projection to identify future needs and future business potentials that may be met by the added functionality. If available, records of past requests for proposals (RFPs) not responded to or proposals rejected by the client may be excellent sources of information on missed opportunities which may be due, at least in part, to lacking functionality.

On the other hand, it is unrealistic to plan for too long a time horizon, for two reasons. Internally, long-run business opportunities and directions become too diffuse for influencing specific tool selections. Externally, the tool vendor market is probably changing faster than the SME's business environment, making long-range projection difficult. New functionalities asked for by a substantial segment of a tool vendor's users tend to become available in subsequent releases and versions. Mergers and teaming arrangements among vendors bring functionalities previously available only in separate tools into one environment. New technical developments, and the software industry's responses to them, make entirely new tools available to the SME.

In summary, the SME needs to develop a precise list of needed and desired functional capabilities for a new tool, derived from an analysis of current and expected future needs, but tempered by a judicious evaluation of potential changes both in the SME's business and in the software vendor industry. Candidate tools can then be evaluated against such a list.

Interoperability. The major theme of this report is:

> *Advanced Engineering Environments (AEEs), through which people and tools can effectively interoperate in the delivery of engineering products and services, are becoming feasible even for the smallest SME.*

Today, interoperability ranks a close second to functionality as a tool selection criterion. The day of the independent tool is long gone. SMEs are no longer willing to manually copy the output of one tool to serve as the input of the next tool – an effort that is neither productive nor cost-effective, albeit a practice all too common in the early days of computer use.

Interoperability among tools can be achieved in a variety of ways. Among the modes of interoperation discussed in the previous report and identified in the two assessment tools presented in Chapter 2 are:

- Tools communicating in a 'native mode';
- Tools interfaced through common files (STEP or other standard or neutral format);
- Tools interfaced through a shared database;
- Tools integrated (by their vendors) over a common shared database; and
- One tool operates within another.

The most important criterion in selecting among competing tools on the basis of interoperability considerations is the degree to which the interoperation mode offered by the tools supports the organization's overall product development process. A pair of illustrative comparisons will make this point clearer. On the one hand, assume that an SME separates the concerns of its design and manufacturing divisions to the point where for each product only one data transfer takes place from the CAD system to the CAM system at the completion of the design process, in waterfall fashion. In this case, even the slowest transfer mode is satisfactory. On the other hand, assume a different organizational structure, where the manufacturing division enters early in the design process and emergent designs are frequently sent to the manufacturing division for evaluation and feedback on manufacturability issues. In this case, the SME needs a data transfer mode and a communication interface that reduce the delays in the feedback loop to a minimum. Such intimate interaction can be achieved if there are no translations to be made, so that two engineers can look at the same model at the same time and discuss costs and changes. A similar distinction applies to CAE tools used once per product for final verification vs CAE tools iteratively used in the design process for frequent performance evaluation or optimization of the product as it is being designed.

When interoperation becomes intimate, as in the second alternative of the above illustrations, a second interoperability criterion emerges. In the waterfall CAD/CAM and CAD/CAE scenarios, the two sets of specialists that use the two tools interact so seldom that it does not matter much whether the models within the respective tools are understandable to the other discipline or not. Human interaction is primarily face-to-face or through drawings, plots, web pages or e-mail messages. All of these "data transfer modes" are much more flexible and redundant than direct transfers between tools. As the interaction becomes more intimate, it becomes increasingly important that the computer-based models used by the tools be understandable, to some degree at least, to the specialists in the interacting disciplines. Otherwise, the potential offered by rapid iterations in the design process will not be fully exploited as people have to slow down to mentally translate strange models into their own terms.

Whenever translation between computer-based models is involved, regardless of the transfer rate, a third interoperability criterion emerges, that of translation fidelity. Fidelity is a measure of the extent to which the tool receiving the translated data can construct a complete and faithful computer-based model for its own purposes, without loss of information. It is a complex function of the representations used by the interoperating tools and the comprehensiveness of the translator program. Loss of fidelity in transfer can have serious technical and financial implications, and should be extensively tested as part of the tool selection process.

Usability. It is a truism that a tool must be usable in order to be effectively, even enthusiastically, used. At the basic level, the layout of the interfaces, the function and location of various control features and the familiarity of the "look and feel" of the tool are usability criteria for every computer-based tool. Software vendors and universities have gained increased understanding of human-computer interaction issues and principles and this knowledge has significantly improved tool interfaces.

For technical tools, such as CAD, CAM and CAE tools, there are three additional usability criteria. First, users insist that the tool interface and the way of specifying actions by the tool be "intuitive." This imprecise term means that the consequences of specifying an action to the tool, and the responses displayed by the tool, and should be what the user expects, based on his/her education, training, and experience with previous tools of the same class.

Second, technical users expect that the tool be "transparent" to some extent, and not a "black box." A black box does not reveal anything about its inner workings; input goes in and output comes out, with the user left totally in the dark on how the latter was derived from the former. In contrast, a transparent program makes some attempt to explain its reasoning in a terminology familiar to the user. Full transparency is not easy to achieve, and may even be counterproductive. The user does not expect a CAE tool implementing finite element analysis to display all of its intermediate steps; but he/she has the right to expect occasional status messages ("assembling stiffness matrix," "solving equations," etc.) and query capabilities for intermediate results or checks. The third criterion, customizability, is discussed below.

Expandability. The last set of technical selection criteria deals with the manner in which the tool's functional capabilities may be expanded in the future. Expandability has two aspects: external and internal expandability. External expandability pertains to what may be obtained from the vendor: the expanded functional capabilities that may be added in the future when needed, and what the expansion entails. As indicated repeatedly in these two reports, many vendors provide both light-duty and heavy-duty versions of their tools as a means of expanding the tool's scope. In other cases, vendors or third-party suppliers provide add-ons to increase tool functionalities. The evaluation criteria for external expandability thus need to address three distinct issues: (1) what expanded capabilities are available; (2) what effort does it take to install the needed expansion(s); and (3) what kind of compatibility is maintained between the original and expanded versions of the tool.

Internal expandability deals with the changes that individual users, or the SME organization on the whole, may make to expand the usability or utility of the tool. This type of expansion is often referred to as "customization." For example, CAD tools may be customized by inserting parts libraries, scripts for defining custom entities, linkages to databases of non-geometric attributes, etc. Obviously, some classes of tools are inherently more customizable than others. At one extreme, spreadsheets are eminently customizable, as they support essentially any operation on tabular data. On the other hand, production-oriented CAE tools, in contrast to research tools, can not be expected to be highly customizable. It is important to establish in advance of tool selection the kinds of customization, compatible with the nature of the tool, that is needed or desired, and evaluate candidate tools against this list.

4.1.2 Service criteria

Vendor support & training. The single most important non-technical criterion for selecting a tool is the extent of the vendor's commitment to provide support and training prior, during and after installation of the selected tool. Such support and training are essential for all levels of tools. Clerical and data entry personnel need to be trained and supported so as to be able to execute all tasks within their domain with dispatch and confidence; they can not be expected to experiment with alternate approaches when a tool malfunction occurs. Engineers and designers, through their technical background, are more willing to experiment and even to try to "break the tool." On the other hand, they need extensive training, opportunities to experiment, and a wide set of examples to work on until they have "internalized" the tool to the point where they are willing to make professional decisions based on the results from the tool. The need for this type of advanced professional training cannot be overemphasized.

The potential vendor's capability to deliver the kinds of support described above needs to be clearly determined, and its track record explored. The relationship between users and vendors is increasingly becoming one of partnering. The SME needs to recognize that its continued capability to profitably deliver products or services is increasingly dependent on an outside entity, the tool vendor. With the startups, acquisitions and mergers taking place in the software industry, the future of this partnership is at least as difficult to predict as the future technical needs. Nevertheless, predictions need to be made and candidates need to be evaluated on this basis as well.

Staff interests. Computer-based tools are not deployed in a vacuum. Continued and effective use of a tool requires that the personnel using it be actively involved in its acquisition, installation, use, and upgrading, when needed. Users need to feel that they are empowered by the tool, that their professional stature is raised and that their performance is qualitatively and quantitatively improved. The quality and level of tools available is increasingly becoming a staff retention issue, after having been a staff recruiting issue for some time.

It is difficult to give crisp and precise selection criteria for staff interests. The best way to incorporate staff interest concerns in the selection and evaluation process is to have a broad segment of the intended user population, from senior members to novices, participate actively in the process.

Cost. Cost is an obvious selection criterion, but seldom is it a discriminating criterion for selection among candidate tools of the same class and with roughly the same set of capabilities. This is because of two reasons. First, the software market tends to self-calibrate itself, so that comparable tools have comparable prices, with only occasional exceptions. Second, the internal costs for staff training, process adjustment, etc., will essentially be the same for any tool in its class.

In developing the costs associated with the installation of a new tool, it is important that the development be comprehensive. Out-of-pocket costs, including the cost of purchasing of the tool, any modifications needed to the previous environment, additional space, furniture, etc., even the direct costs of initial staff training, are generally easy to obtain. Many organizations simply stop with such a list. What tends to be forgotten is the cost of providing the affected staff with time and resources to study, explore, and experiment with a tool until each staff member feels fully competent to use the tool and make professional choices and recommendations based on the results produced by the tool. Organizations that don't plan and budget for these costs tend not to provide the released learning time needed. The affected staff has to bootleg the learning into other tasks, or it is forced to make decisions they feel they are not qualified to make.

4.2 Selection resources

It is clear from the presentation above that many categories of decision-making are combined in developing a set of tool selection criteria and then selecting from among the candidates identified. There are several categories of resources that an SME can tap to assist its decision making, briefly summarized below.

Web-based comparisons. In the early days of computing in engineering and manufacturing, user groups developed around hardware platforms and major tools (then simply called programs). The advice and assistance provided by these peer groups were enormously helpful in fostering the culture of computer use among the pioneering organizations. Today, advice and counsel of trusted peers is still very valuable. With the spread of computing, the peer community has greatly expanded, and peer group organizations have largely disappeared. On the other hand, IT networks and the World Wide Net have brought communities together. There are numerous forums, chat rooms, on-line newsletters, etc., with subject matters relevant to every aspect of engineering environments and their constituent tools. Access to a few of these sites can provide useful input to tool selection.

Magazines, journals. Most technical journals dealing with engineering and manufacturing, particularly the trade magazines distributed free to qualified subscribers covering these fields, provide two kinds of features germane to tool selection. First, many of these publications provide, on a periodic basis, tabular summaries and comparisons of capabilities of classes of tools. Software vendors make sure that their products appear in these summaries, so that the coverage of these summaries is generally quite complete. A recent summary table for a particular class of tools is valuable for the SME in: (1) identifying some of the major tool evaluation and selection criteria; (2) identifying potential vendors; and (3) pruning the list of candidates to consider in detail by eliminating those with tabulated capabilities outside the intended envelope. Second, most publications provide occasional software reviews or feature articles describing, in some depth, capabilities of a new or significantly revised tool. Good reviews in this category, particularly the ones written in the first person, provide a vivid picture of the strengths and weaknesses of the tools described. Reviews of the top candidate tools in a selection process can add significantly to their evaluation.

Trade shows. The opportunity to "check the teeth" or "kick the tires" of candidate purchases has traditionally been part of the selection process. The equivalent of these rituals for computer-based tools is the trade show. These are either stand-alone events or attached to other professional or trade group meetings. Here, booths display the newest versions of tools, and ancillary events provide short courses, discussion and question-and-answer sessions, etc. At such shows, SMEs can gather just about any kind of information on tools of interest, as well as contacts with many sources of further information. The presence of so many competing products in one place can be overwhelming, and two points of advice are in order. First, you have to go to such shows prepared. The amount of information you can glean by stopping at a booth is limited; furthermore, the interface features of all tools in a given category are today so "homogenized" that a brief look will not identify any differentiating characteristic among the tools. Therefore, you have to have made some preliminary pruning and you have to resolve to examine in some detail no more than a handful of the top candidates. Second, you have to be ready for surprises. A new tool or new feature of an existing tool, first unveiled at the trade show, may counteract even the best preparation and alter the list or previous ranking of candidates. A good strategy to follow is to watch for booths surrounded by large crowds and to check these out for relevance.

Consulting Services. Finally, there is a whole range of consulting services that an SME may purchase to assist it in the tool selection and evaluation process. These services range from the preparation of selection criteria and ranking of candidate tools accordingly to full tool selection, installation and the necessary staff training.

4.3 Summary

Technical criteria for selecting a tool to be added to the SME's engineering environments include:

- functional capabilities of the tool, based on current and expected future needs, but taking into account potential changes in business and software;
- interoperability appropriate to support the organization's overall product development process;
- usability, particularly the extent to which the tool is deemed "intuitive" and reasonably "transparent" by its potential users; and
- expandability, in terms of tool features that may be subsequently added and customization that users may apply to extend the tool's capabilities.

Service criteria to consider include:

- extent of support and training provided by the vendor;
- match between the interests of the affected users and the tool's capabilities; and
- cost, including out-of-pocket costs for purchase, installation and training, and internal costs for adequate user experimentation and learning.

The resources available to assist in making tool selections include:

- web-based sources such as forums and chat rooms;
- summary tabular comparisons of tools and feature articles on specific tools in professional and trade publications;
- trade shows demonstrating the tools; and
- consulting services.

5 AEE Technology Adoption

A typical technology installation process contains steps such as choosing the technology product, acquiring it, physically installing it in the work environment, and training operators in how to use the system (which may or may not be related to how they want to use the system in their environment). While all of these actions are necessary to incorporate a new technology into an SME, they are not sufficient, and will not produce the desired benefits for the SME; not until he ~~adopts~~ the technology and ~~integrates~~ it into his operation.

Technology adoption is achieved when the people who need to use the new technology

- are aware of the presence and the status of the technology;
- have appropriate access to it;
- are trained to use it;
- get support for using it; and
- actually *do* use it to support their work tasks [Garcia 02].

5.1 The Adoption Challenge

The benefits of AEE adoption by SMEs are manifold (see Volume I, Chapter 3), and yet many SMEs remain reluctant to adopt advanced software-based technologies. In many cases, this reluctance may be traced to prior experience with unsuccessful software adoptions.

Technology adoption is rarely achieved without challenge. As a new technology is introduced, the work of the organization must continue. As such, the adoption effort is superimposed upon normal workday activities. As the technology is introduced, the organization may face challenges such as:

- Diverting critical resources from ongoing production activities to technology analysis and selection activities;
- Diverting critical production resources for training;
- Modifying time-tested work practices to utilize the capabilities of the new technology;
- Changing the skill sets within the organization to accommodate the installation, utilization, and support of the new technology; or
- Overcoming the common human resistance to change.

These challenges can impact the performance of the organization, as illustrated in Figure 3 [Garcia 02].

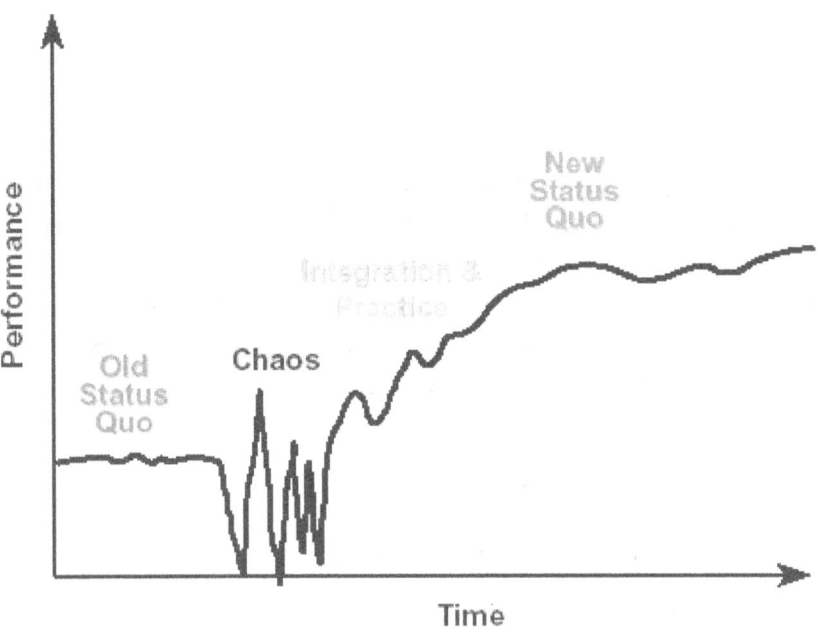

Figure 3: The Impact of Technology Adoption

Change is difficult for people and for organizations. Technology adoption is synonymous with change: a change of technology; a change of users skills; a change of the way work is done.

> "There is nothing more difficult to take in hand, more perilous to conduct or more uncertain in its success than to take the lead in the introduction of a new order of things."
>
> Niccolò Machiavelli: *The Prince*, 1532

One method of reducing the fear of change and improving the probability of successful technology adoption is to implement adoption via an orderly process; one that:

- Exposes the need for change to all of the stakeholders;
- Involves the stakeholders in the development of the adoption plan;
- Defines the future state of the organization after the adoption;
- Maps the path from the present state to the future state for all to see; and
- Minimizes the risks of technology adoption.

5.2 The Technology Adoption Process

The goal of technology adoption is operations and process improvement, an area of study addressed vigorously since the 1920s. Many models exist for process improvement, among them the Shewart Cycle and the Initiate, Diagnose, Establish, Act, and Leverage (IDEAL) model not elaborated here [McFeeley 96]. The remainder of this chapter will discuss technology adoption using a modification of the Shewart Cycle.

The Shewart Cycle is an early model developed for process improvement [Shewart 39]. Although its creation predates the existence of software, it remains fully applicable to software-based technology adoption today. The Shewart Cycle was later articulated by W. Edwards Demming as "Plan, Do, Study, Act" [Demming 82]. Presently, the universally accepted nomenclature is "Plan, Do, Check, Act" or PDCA. These four activities comprise a method for achieving continuous process improvement. The four steps are described as follows:

Plan: Define a plan for a new process or improvements to an existing process. This plan must include monitoring and data collection methods as well as performance objectives.

Do: Implement the proposed changes, on a small scale, perhaps as a pilot project. Obtain performance measurements before, during and after the pilot.

Check: Study the measurements collected during the pilot. Analyze the results to identify and understand failures and successes. Determine the "lessons learned" during the pilot.

Act: Act to apply the conclusions of the prior analysis.
NOTE: Implementation, cycle restart, or abandonment are all acceptable actions.

This process is illustrated in Figure 4.

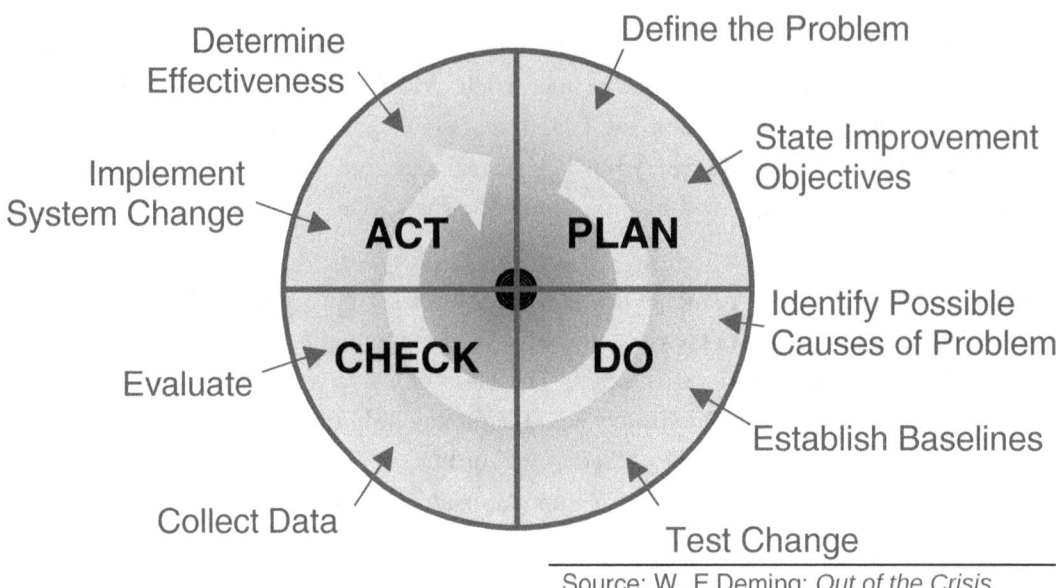

Figure 4: Shewart Cycle

A derivative of this process specifically adapted for technology adoption is shown in Figure 5 [Garcia 03]. This figure represents one PDCA cycle.

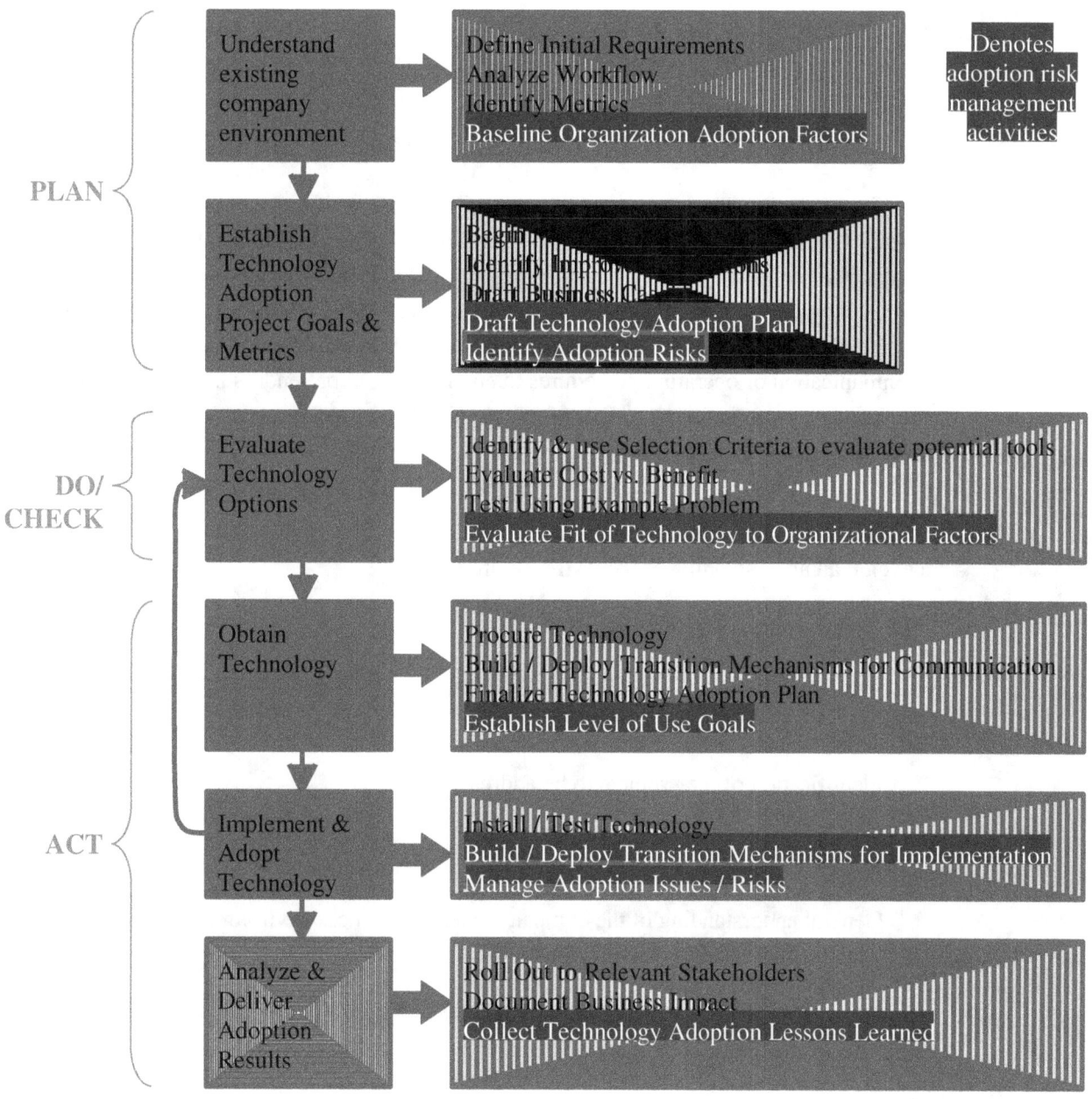

Figure 5: Technology Adoption Process

The six phases of this technology adoption process are:

1. Understand existing company environment;
2. Establish Technology Adoption Project Goals & Metrics;
3. Evaluate   Technology Options;
4. Obtain   Technology;

5. Implement & Adopt Technology; and
6. Analyze & Deliver Adoption Results.

These phases are described in detail in the following sections.

5.2.1 Understand existing company environment

The Technology Adoption Process of Figure 5 focuses significant attention on the planning phase of technology adoption. Research has shown that this is of critical importance to SMEs because they tend to rely on informal rules and infrastructures, often depending on "tribal" knowledge (i.e., a mixture of experience and expertise transmitted from one worker to another). Rules or infrastructures for decision-making are often fuzzy. Documentation and communication of operating procedures to employees and/or vendors is often unclear. Consequently, predicting the effect of advanced software technologies upon the organization may be difficult [Estrin 03].

To address these limitations the Technology Adoption Process uses a four-step method to develop an understanding of the SMEs environment.

1. Define Initial Requirements

To initiate a technology adoption, the stakeholders within the organization first define business and technical needs to be addressed. This is done based upon:

- Identification of the problem to be addressed;
- Company vision;
- Stakeholder input; and
- General understanding of the company external business environment (competitors, market drivers, technology advances, etc.).

This effort results in a statement of requirements and the assignment of resources for further effort on the project

2. Analyze workflow

An early task of the project team is to document the "As-Is" business processes in place at the company relevant to the problem being addressed. This involves identifying process steps and work products related to the problem. Research is based upon company process documents, technology improvement requirements, company employee interviews, etc. The effort results in a description of the current company process(es). This is useful in understanding the role of the potential technology adoption, and also provides a baseline against which improvement can be measured.

3. Identify Metrics

Performance measures must be defined to enable monitoring of progress and recognition of success. For SMEs, this is typically tied to how managers communicate company performance to owners / executives. Metrics are typically derived from the improvement requirements, the "As-Is" process description, input from stakeholders, etc.

4. Baseline Organizational Adoption Factors

Technology adoption can be viewed as a convergence of technology, organization, and people; i.e., people operating within an organization to incorporate a new technology into their daily routine. The sociological and organizational context of the technology adoption is often a key factor in developing a technology adoption plan defining the application of assets. [Adler 90] and the roles of the stakeholders [Gladwell 02].

5.2.2 Establish Technology Adoption Project Goals & Metrics

The second phase of the technology adoption process is the development of an adoption plan. The focus of these activities is to ensure that the technology adoption supports the business objectives of the company, and to ensure that the results of the adoption are measurable. While this focus is important in any organization, large or small, it is critical for the SME. While there are many reasons for the technology adoption failures among SMEs; failure to link technology adoption to key business objectives is one of the more common [Buhman 03]. In creating this linkage, the SME must understand current business practices, understand current operations processes, understand the company's strategic goals, and understand the company's desired future state.

Technology adoption is frequently both a capital expense and a workforce development effort. For SMEs, both of these are a challenge:

- capital investment is difficult due to the often limited financial resources of the SME; and
- workforce development is difficult due to both the expense of training and the "out-of-the-office" time of key employees.

For these and other reasons, SMEs are often very conservative when investing in technology adoption, and demand a "bulletproof" business case before proceeding.

The technology adoption process builds this business case in five steps.

1. Begin Metrics Collection

Begin collection of metrics to establish a baseline against which future improvements can be evaluated. This activity may also involve calculating measures from a previous time period using available data. This activity is often the first time that an SME has obtained an in-depth view of the process to be improved. This clear view of an inefficient and/or ineffective process will often serve as a stimulus to reinforce the process improvement activity.

2. Identify Improvement Options

Identify multiple improvement options that meet the Improvement requirements. Many ideas may come from company employees at all levels and some improvement options may or may not include adoption of new technologies. Ideas that involve a technology should remain general (example: an option should say adopt 3-D CAD vs. arbitrarily specifying a 3-D CAD tool vendor).

3. Create a Draft a Business Case

Evaluate the improvement options to determine both the total cost (e.g., capital investment, training costs, sustainment costs, licensing) of technology adoption and the expected return.Estimate and analyze cash flow, a subject critical to many SMEs. Create and distribute a draft business case summarizing this information, along with measures, approach and rationale.

4. Create a Draft Technology Adoption Plan

Create a draft technology adoption plan. This plan should include:

- Delineation of stakeholder expectations;
- Adoption process schedule and milestones;
- Adoption process staffing;
- Communication processes; and
- Constraints (e.g., time, money, staff availability, user skills) imposed upon the adoption process.

5. Identify Adoption Risks

Define detailed adoption context by identifying barriers and risks to adoption, including people and organizational constraints. For example, identify Adler weaknesses is term of organizational assets and appropriateness of roles [Gladwell 02].

5.2.3 Evaluate Technology Options

The third phase of the technology adoption process is the evaluation and selection of the appropriate technology.

1. Cost Benefit Analysis

Document the costs and benefits of each of the improvement options, including initial and recurring costs as well as both tangible and strategic benefits. Prioritize the improvement options based upon the draft business case and the draft technology adoption plan. Select the best option based upon current understanding. Selecting multiple improvements at one time should be avoided to avoid the chaos zone mentioned in [Weinberg 97].

2. Identify and Use Selection Criteria To Evaluate Potential Tools

If the selected improvement option involves adoption of a new technology, identify selection criteria and begin an evaluation process to select a specific product that supports the technology.

3. Test Using a Sample Problem

Identify a model problem to test high risk aspects of the proposed solution. This is also important to validate vendor claims and identify changes needed to integrate the technology into the future-state business process.

4. Evaluate Fit of Technology to Organization Factors

Compare the desired technology adoption to the capabilities of the organization and consider the risks and supporting context. In addition, this activity can point to likely adoption patterns and provide insight into the plan and schedule of the technology adoption.

5.2.4 Obtain Technology

1. Procure Technology

Finalize agreement with the technology vendor to purchase the selected product and any necessary training and/or consulting support.

2. Establish Level of Use Goals

Define level of use goals for the technology. Define these goals not just as the magnitude of technology use, but in terms of roles and process steps. For example, set objectives for when a 3D CAD tool is used within a development process (like the proposal step) and which roles (e.g., Engineer, Designer) use the tool.

3. Finalize Technology Adoption Plan

Update and finalize the draft technology adoption plan to include all activities required to move from the As-Is to the To-Be business processes.

4. Build/Deploy Transition Mechanisms for Communication

Create the necessary communication items (e.g., progress reports, training plans) and distribute to the stakeholders.

5.2.5 Implement & Adopt Technology

1. Install Technology

Insert the technology/tool into the company business using the To-Be business process.

2. Build/Deploy Transition Mechanisms for Implementation

Provide all transition items for implementing the To-Be process as described in the technology adoption plan, including training, consulting, and transition monitoring and metrics collection.

3. Manage Adoption Issues and Risks

Actively and regularly monitor the adoption risks and context because the context is always changing. This includes observing progress against metrics and observing indicators of technology adoption bottlenecks.

5.2.6 Analyze & Deliver Adoption Results

This final phase of the technology adoption process is focused upon institutionalizing the adoption process within the SME. Part of this process involves building support for the future technology adoption efforts by publicizing the success of past and current efforts.

1. Document the Business Impact

Summarize the business impact in terms of tangible and strategic benefits. This also includes asking employees about impact throughout the implementation.

2. Collect Technology Adoption Lessons Learned

Examine how the organization (including people, assets and context) performed in the technology adoption and identify ways to improve in technology adoption. For example, did the organization meet the technology level of use objectives as well as the business objectives?

3. Roll Out to Relevant Stakeholders

Summarize the business case and communicate it to the company stakeholders, both internal and external.

6 Summary

The intent of this two-volume report has been to build awareness of the AEE concept, and to provide guidance to SMEs considering the adoption of AEE technology. This chapter summarizes key concepts presented in these reports.

AEEs are integrated toolsets that enhance the productivity of participants in the product development and production processes. They are defined as computational and communications systems that can create virtual and/or distributed environments functioning to link researchers, technologists, designers, manufacturers, suppliers, and customers.

In these reports, AEEs are classified in three levels. A **Basic AEE**, well within the reach of most SMEs, consists of only two well-matched COTS software products providing design and analysis capabilities. An **Intermediate AEE** is also suitable for implementation by SMEs. While also composed of COTS design and analysis products, it provides higher levels of data integration and interoperability than the Basic AEE. A **Comprehensive AEE** providing total data sharing and interoperability with both technical and non-technical IT systems throughout the organization does not yet exist. This description serves primarily as a roadmap of future developments and standardization efforts at the component interfaces.

An AEE enables an SME to utilize more efficient and more effective design processes. Without the benefits of an AEE, an SME is forced to design a product based upon experience and limited manual analysis processes. True functionality of the product is not known until it is built and tested. With an AEE, the SME may analyze and predict end-product performance while still in the design stage, enabling optimization of the design. As the SME migrates to more advanced AEEs, this prediction and optimization process becomes faster, easier, and more effective.

Some of the benefits an AEE can offer to an SME include:

- **Reductions in product development time** - AEEs encourage and support design reuse and parametric design processes, eliminate redundant efforts during the design process, encourage functional analysis earlier in the design cycle, and encourage collaboration among designers, engineers, manufacturers, suppliers, and customers.
- **Reductions in production time** - AEEs enable design optimization by eliminating overly conservative assumptions. They minimize component inventories by encouraging

and supporting design reuse. They also enable process designers to simulate and optimize manufacturing processes during the product design stage.

- **Improvements in product quality** - AEEs encourage and support functional analysis and simulation prior to manufacturing. They enable design optimization by evaluating the impacts of design trade-offs. They validate design performance prior to manufacturing. They encourage and support multi-disciplinary collaboration among designers, engineers, manufacturers, suppliers, and customers. They encourage and support reuse of existing successful designs. They enable the use of advanced simulation techniques such as Monte Carlo and Taguchi methods.

- **Reductions of product cost** – AEEs enable design optimization, eliminating costly, overly conservative assumptions. They encourage and support design reuse, minimizing the required component inventory. They enable process designers to simulate and optimize manufacturing processes during the product design stage. They can also reduce maintenance costs, spare parts inventories, warranty costs, field changes and upgrades. That is, doing it right up front has enormous impact on the cost of the product over its life-cycle, to both the producer and the customer.

- **Reductions of product development cost** – AEEs create reductions in the product development schedule (see above). They detect functional problems earlier in the design process, when they are less costly to fix. They encourage and support design experiments early in the design process.

- **Improved market agility** – AEEs enable the SME to cope with rapidly changing global market demands and competitive environments.

- **Improved communications** - AEEs provide a means of improving communications with both customers and suppliers, strengthening the supply chain.

The process of adopting an AEE begins with the evaluation of the current state of the SME. The SME must determine the current mix of products, degree of internal integration, current level of computer use, current skill and knowledge levels, and perceived problems. This is followed by the definition of a goal state, including the desired future state of the SME, and the strategies to overcome the perceived problems. From this information, the SME may determine the scope and level of the AEE to be adopted, leading to the specification of the requirements for the components of the AEE, and the selection of the COTS AEE components satisfying these requirements.

The components described in this report cover a wide range of characteristics necessary to implement AEEs. A single SME does not need to implement all of these components, but needs to carefully identify the components necessary for its specific needs, the planned interactions between them, and the range of COTS products implementing the needed components. A carefully prepared plan will lead to successful expansion of the AEE with additional components in the future. These reports provide a tool to assist the SME in defining the value of an AEE to its operation, and a tool to assist the SME in identifying the AEE content.

Many SMEs evolve or migrate from lower AEE levels to higher ones. As new tools and capabilities are introduced into a company's support environment on an incremental and opportunistic basis, the SME must recognize the shifts in operational modes and philosophies will be needed to gain the full benefit of the AEE.

In choosing a tool to be added to the SME's engineering environments, the SME must consider the functional capabilities of the tool, based on current and expected future needs; however, the SME must also take into account potential changes in business and software. The SME must also ensure that the tool's interoperability is appropriate to support the organization's overall product development process. Usability, particularly the extent to which the tool is deemed "intuitive" and reasonably "transparent" by its potential users, is a key factor in tool selection. Expandability, in terms of tool features that may be subsequently added and customization that users may apply to extend the tool's capabilities, should also be considered.

Selection, procurement, and installation of a new technology will not produce the desired benefits for the SME; not until it adopts the technology and integrates it into its operation. Only when the SME's staff is aware of the technology, has access to it, is trained to use it, gets support for using it, and actually DOES use it will the benefits accrue to the SME.

A six-step process based upon the Shewart Cycle is described. The process has been used successfully to introduce advanced technologies into SMEs and consists of the following steps:

1. Understand existing company environment;
2. Establish technology adoption project goals & metrics;
3. Evaluate technology options;
4. Obtain technology;
5. Implement and adopt technology;
6. Analyze and deliver adoption results.

References/Bibliography

[Adler 90]	Adler, P.S. & Shenhar, A. (1990). *Adapting your technological base: The organizational challenge.* Sloan Management Review 32(1), 25-37.
[Buhman 03]	Buhman, C., *Evaluating and Selecting Software in a Smaller Manufacturing Company.* Pittsburgh, PA: Dynamic Business Volume 58, Number 2 (2003): 18-19
[Demming 82]	Demming, W. E., *Out of the Crisis.* Cambridge, MA: MIT, Center for Advanced Engineering Study, 1982
[Estrin 03]	Estrin, L., Foreman, T. & Garcia, S., *Overcoming Barriers to Technology Adoption in Small Manufacturing Enterprises (SMEs).* Pittsburgh, PA: Software Engineering Institute, Carnegie Mellon University, 2003 available at http://www.sei.cmu.edu/tide/publications/
[Garcia 02]	Garcia, S. & Seeley, G., *Beyond Technology Installation: Realizing Better Return from Your IT Investments.* Pittsburgh, PA: Software Engineering Institute, Carnegie Mellon University, 2002 available at http://www.sei.cmu.edu/tide/publications/
[Garcia 03]	Garcia, S. & Estrin, L., *Managed Technology Adoption: Part I* *The Way to Realize Better Return From Your IT Investments.* Pittsburgh, PA: Dynamic Business Volume 58, Number 7 (2003): 16-17
[Gladwell 02]	Gladwell, M., *The Tipping Point: How Little Things Can Make a Big Difference.* Boston, MA: Back Bay Books; 2002 ISBN: 0316346624
[McFeeley 96]	McFeeley, R., *IDEAL: A User's Guide for Software Process Improvement.* Pittsburgh, PA Software Engineering Institute, Carnegie Mellon University, 1996 available at http://www.sei.cmu.edu/publications/documents/96.reports/96.hb.001.html
[Shewart 39]	Shewart, W., *Statistical Method from the Viewpoint of Quality Control.* Washington, DC: Department of Agriculture, 1939 and Dover, 1968

Acronyms

2D	Two-Dimensional
3D	Three-Dimensional
AEE	Advanced Engineering Environment
ASP	Application Service Provider
CAD	Computer Aided Design
CAE	Computer Aided Engineering
CAM	Computer-Aided Manufacturing
CMU	Carnegie Mellon University
COTS	Commercial Off-the-Shelf
FEA	Finite Element Analysis
NC	Numerical Control
NIST	National Institute of Standards and Technology
OMG	Object Management Group
PDM	Product Data Management
RFQ	Request for Quotation
SEI	Software Engineering Institute
SME	Small Manufacturing Enterprise
STEP	STandard for the Exchange of Product model data
TIDE	Technology Insertion, Demonstration, and Evaluation

REPORT DOCUMENTATION PAGE

Form Approved
OMB No. 0704-0188

Public reporting burden for this collection of information is estimated to average 1 hour per response, including the time for reviewing instructions, searching existing data sources, gathering and maintaining the data needed, and completing and reviewing the collection of information. Send comments regarding this burden estimate or any other aspect of this collection of information, including suggestions for reducing this burden, to Washington Headquarters Services, Directorate for information Operations and Reports, 1215 Jefferson Davis Highway, Suite 1204, Arlington, VA 22202-4302, and to the Office of Management and Budget, Paperwork Reduction Project (0704-0188), Washington, DC 20503.

1. AGENCY USE ONLY (Leave Blank)	2. REPORT DATE	3. REPORT TYPE AND DATES COVERED Final
4. TITLE AND SUBTITLE Advanced Engineering Environments for Small Manufacturing Enterprises: Volume II		5. FUNDING NUMBERS F19628-00-C-0003
6. AUTHOR(S) Steven J. Fenves (NIST), Joseph P. Elm (SEI), John E. Robert (SEI), Ram D. Sriram (NIST)		
7. PERFORMING ORGANIZATION NAME(S) AND ADDRESS(ES) Software Engineering Institute Carnegie Mellon University Pittsburgh, PA 15213		8. PERFORMING ORGANIZATION REPORT NUMBER
9. SPONSORING/MONITORING AGENCY NAME(S) AND ADDRESS(ES) HQ ESC/XPK 5 Eglin Street Hanscom AFB, MA 01731-2116		10. SPONSORING/MONITORING AGENCY REPORT NUMBER
11. SUPPLEMENTARY NOTES		
12A DISTRIBUTION/AVAILABILITY STATEMENT Unclassified/Unlimited, DTIC, NTIS		12B DISTRIBUTION CODE

13. ABSTRACT (MAXIMUM 200 WORDS)

To assist the Small Manufacturing Enterprise (SME) in adopting Advanced Engineering Environments (AEEs), this report provides two self-assessment tools; the Self Assessment Tool for Engineering Environments (SAT-EE) to assist an SME in assessing the adequacy of the current computing support environment in handling technical tasks, and the Self Assessment Tool for Engineering Tool Capabilities (SAT-ETC) to collect the needs and desires of the operation, and map them to the capabilities of specific classes of CAD and CAE tools. These tools help the SME evolve from lower AEE levels to higher ones, shifting operational modes and philosophies to gain the full benefit of the AEE.

In choosing an AEE component the SME must consider the tool's functional capabilities, interoperability, usability, and expandability. Selection, procurement, and installation of a new technology must be followed by tool adoption to integrate it into the SMEs operation. A six-step process based upon the Shewart Cycle has been used successfully to introduce advanced technologies into SMEs

14. SUBJECT TERMS			15. NUMBER OF PAGES
16. PRICE CODE			
17. SECURITY CLASSIFICATION OF REPORT Unclassified	18. SECURITY CLASSIFICATION OF THIS PAGE Unclassified	19. SECURITY CLASSIFICATION OF ABSTRACT Unclassified	20. LIMITATION OF ABSTRACT UL

www.ingramcontent.com/pod-product-compliance
Lightning Source LLC
Chambersburg PA
CBHW081737170526
45167CB00009B/3847